SQUIRM The Earthworm

Dedication

I dedicate this book to my nieces and nephews.

Acknowledgment

I would like to thank Miriam Glover Marketing for her professional editorial skills and Alexander Ray Watkins Jr., for his brilliant artwork throughout this entire process.

SQUIRM
THE EARTHWORM
A SCIENCE RHYMING BOOK

BY AQUILLA DANIELS

Hi, my name is Squirm.
I am an earthworm.

I am not a creepy icky creature.
I'm really shy by nature.

I have tons of cool features
Like…

Covered in slime; I look this way all
the time.

Oxygen melts in my skin, so I can
take air in.

Shaped with no bones, my tummy constantly moans.

I have no teeth, but I love to eat! Fruits and veggies, but please no meats. I much prefer pasta noodles, salad scraps, banana peels, and paper sheets!

I live underground and when the sun sets,

I crawl to the surface to begin my work. I swallow dead leaves and fertilize all forms of dirt.

"Slurp. Excuse me! I just burped!"

You see, I am a night crawler and a garden's friend.

I tidy up every night and leave no loose ends.

I'm different than most but I have no regrets. "You bet!"

Although I am unable to see or hear, I feel vibrations in the ground when animals come near.

All day strong, my senses are life-long.

Amazingly, earthworms have five hearts and are exceptionally smart!

With no eyes, legs, or arms my friends all ask, "Squirm the Earthworm, how do you move?"

Well, I answer, "It's no easy task!"

But once I start, I have a nice groove,

"Squeeeeeeze then strettttchh muscles tighten then release, tiny hairs grip the dirt and I go, go, go."

Breathe in and out through my slimy skin. The air is so thin.

Back to work I go, there's so much dirt to eat. These burrows that I make are quite a feat! Each tunnel guides raindrops that feed plant roots.

"But all this eating makes me want to...oops!"

I work hard so earth's soil stays nutritious and lean, so farmers can grow healthy green beans. Not just beans, but fruits, and veggies of all kinds. I smile at all the foods you can find.

"Squeeeeeeze then strettttchh muscles tighten then release, tiny hairs grip the dirt and I go, go, go."

Back and forth, my body moves like a broom. Aerating the soil so colorful flowers can bloom.

Life can be scary, with many dangers all around. From flying birds in the sky, to snakes and big-eyed toads on the ground.

But I, Squirm the Earthworm is up to the challenge.

I am careful when crossing the street and try my best not to be their next tasty treat.

In damp places, my skin stays moist and never firm. Too much sunlight makes me burn.

"Squeeeeeeze then strettttchh muscles tighten and release, tiny hairs grip the dirt then I go, go, go."

If you see me moving, fast or slow, please don't SQUASH me with your enormous toe.

It's okay to say, "What's up, Squirm?" as you gently pick me up.

Keep your hands shaped in a cup then lay me in the dirt to do my earthworm work.

Earthworms help the environment to grow and stay strong. Scientists say, "Saving my life can do no wrong."

Together, our beautiful earth is where you and I belong!

The End

www.ingramcontent.com/pod-product-compliance
Lightning Source LLC
Chambersburg PA
CBHW052056190326
41519CB00002BA/244